Regenerative Engineering Society

Held at the 2020 AIChE Annual Meeting

Online
16 – 20 November 2020

ISBN: 978-1-7138-2286-8

Regenerative Engineering Society

Held at the 2020 AIChE Annual Meeting

Online
16 – 20 November 2020

AIChE
The Global Home of Chemical Engineers

Printed from e-media with permission by:

Curran Associates, Inc.
57 Morehouse Lane
Red Hook, NY 12571

Some format issues inherent in the e-media version may also appear in this print version.

Printed with permission by Curran Associates, Inc. (2021)

For permission requests, please contact AIChE
at the address below.

AIChE
120 Wall Street, FL 23
New York, NY 10005-4020

Phone: (800) 242-4363
Fax: (203) 775-5177

www.aiche.org

Additional copies of this publication are available from:

Curran Associates, Inc.
57 Morehouse Lane
Red Hook, NY 12571 USA
Phone: 845-758-0400
Fax: 845-758-2633
Email: curran@proceedings.com
Web: www.proceedings.com

TABLE OF CONTENTS

616639 Additive manufacturing in treatment for bone disorders: Opportunities and challenges

Abstract:

Role of 3D printing (3DP) or additive manufacturing (AM) is becoming important in need of patient matched implants due to lower cost and shorter lead time to manufacture. Additively manufactured components need to be controlled and optimized carefully for their reproducibility, machine to machine part quality variations and process specific material properties. Establishing process property relationships for different AM techniques are vital towards successful implementation of these manufacturing practices in biomedical devices. Hard biomaterials, e.g., calcium phosphate (CaP) ceramics being compositionally similar to the inorganic part of bone, show significant promise towards bone implant applications, in both 3DP tissue engineering scaffolds and surface modified hip and knee implants. Dopant chemistry in CaP plays a vital role in controlling their resorption or degradation kinetics as scaffolds, mechanical strength, and biological properties of resorbable CaPs. In vivo studies show improved osteogenesis, angiogenesis and controlled drug delivery using natural medicinal compounds (NMCs) in these 3DP scaffolds and coatings. These systems show promise for their use in orthopedic and dental devices while eliminating the need for the autografts and the second site surgery for harvesting, as well as improving current hip / knee implant lifetime.

613706 Synergistic Effects of Graphene and Nanofibers Alignment on Muscle Regeneration

Abstract:

Introduction: Skeletal muscle comprises 40-45% of the total human body mass and consists of bundles of aligned muscle fibers. Since current strategies cannot achieve satisfactory results for the treatment of severe muscle injuries, novel tissue regeneration strategies are being developed to address current challenges. The aligned orientation of the scaffold provides an essential topographical cue to mimic the natural structure of muscle tissue and enhance the myoblasts growth and differentiation compared with random fibers. Since muscle is an electrically excitable tissue the use of electroactive biomaterials along with biomimetic aligned nanofibers facilitate electrical signal transfer and myoblast communication, growth, and development. Studies reported the role of biochemical cues to induce myoblast differentiation, elongation, and alignment by switching the culture media from standard growth media (GM) to differentiative media (DM). Here, we successfully demonstrated that highly aligned pristine graphene-based nanofibers can significantly induce myotube formation in standard GM under the condition without electrical stimulation.

Materials and Methods: The scaffolds were fabricated by electrospinning the composite solutions of pristine Graphene nanoplatelets (GnPs) and poly (l-lactic acid) (PLLA) with different amounts of GnPs include 0, 0.5, 1.5, and 2 mg/ml. The micromorphology of the nanofibers was examined via scanning electron microscopy (SEM). The nanofiber diameter, surface functional groups, mechanical properties, and electrical conductivity were evaluated by Image J software, Fourier Transform Infrared spectroscopy (FTIR), Instron machine, and four-point probe device, respectively. To investigate the muscle cells proliferation and differentiation, mouse C2C12 myoblasts were cultured on the scaffolds in standard GM and investigated through live/dead assay, CCK8 assay, and immunostaining with myosin heavy chain (MHC).

Results and Discussion: Based on the results, incorporation of GnPs in PLLA nanofibers significantly improved mechanical properties and enhanced electrical conductivity compared with pure PLLA scaffold while no significant difference was observed in the morphology of the scaffolds. Fig1.A shows that all concentrations of pristine GnPs promoted myoblasts growth and proliferation 7 days after cell culture. Moreover, the significant formation of myotubes on all the scaffolds in standard GM can be observed in Fig1.B. The statistical analysis demonstrated that topographical cue even in standard GM had the highest impact on the length of myotubes while electrical cue in biomimetic DM had the highest effects on myotube fusion and maturation. The combination of topographical cue and electrical cue demonstrated the highest differentiation and maturation grade on the formed myotubes in GM, suggesting that the guidance cues further promoted aligned, long, and multinucleated myotube formation. The myotube length, fusion, and maturation indices of A(GM)-GnP 1.5 were further increased to 411.4 ± 160.8 μm, 70.37 ± 3.044%, and 57.59 ± 7.163%, respectively (Fig 1.C-E). These results showed that the combination of topographical and electrical cues in standard GM was more efficient than a single cue alone in DM for stimulating myoblast differentiation and myotube maturation.

Figure 1. A) Live/dead fluorescent images of the random and aligned scaffolds after 7days (green=viable cells; red=dead cells), Arrows show the main directions of nanofibers alignment. B) Immunofluorescent images of differentiated myotubes on the random and aligned scaffolds after 7 days (MHC=green and nucleus=blue). Quantification of C) Myotube length, D) Fusion index, E)

Maturation index, (* = P ≤ 0.05, ** = P ≤ 0.01, *** = P ≤ 0.001, **** = P ≤ 0.0001; n=5).

Conclusions: The study showed that a combination of electrical and topographical cues can significantly improve the physical properties of the scaffolds. Incorporation of GnPs as an electrical cue into highly aligned nanofibers significantly improved the myoblasts growth and proliferation and induced myotube formation in standard GM without the need for external electrical stimulation.

Acknowledgments: We acknowledge the financial support received from DHHS/NIH/ NIAMS entitled, Regenerative Engineering of Complex Musculoskeletal Tissue and Joints, Grant No.: DP1AR068147.

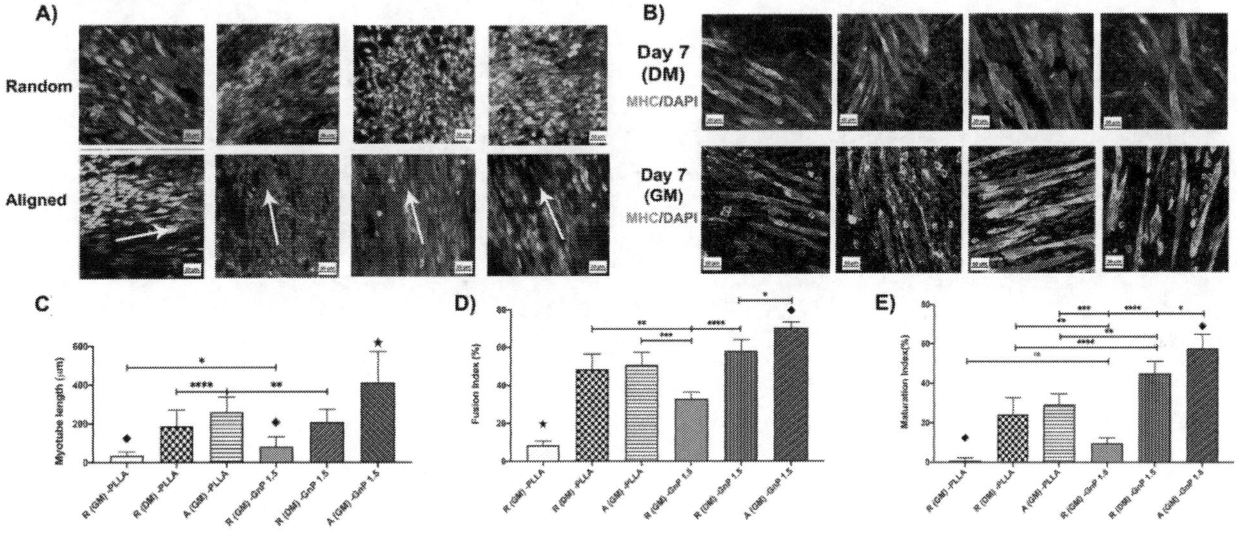

University of Connecticut

UCONN
SCHOOL OF ENGINEERING

RNAi therapy using Janus base nanotubes against cartilage degradation

Yupeng Chen, Ph.D.
Associate Professor
University of Connecticut

Non-covalent Janus-Base Nanotubes (JBNTs)

Janus:
In ancient Roman religion and myth, **Janus** is the god of beginnings, transitions, time, and endings. He is depicted as having two faces, since he looks to the future and to the past.

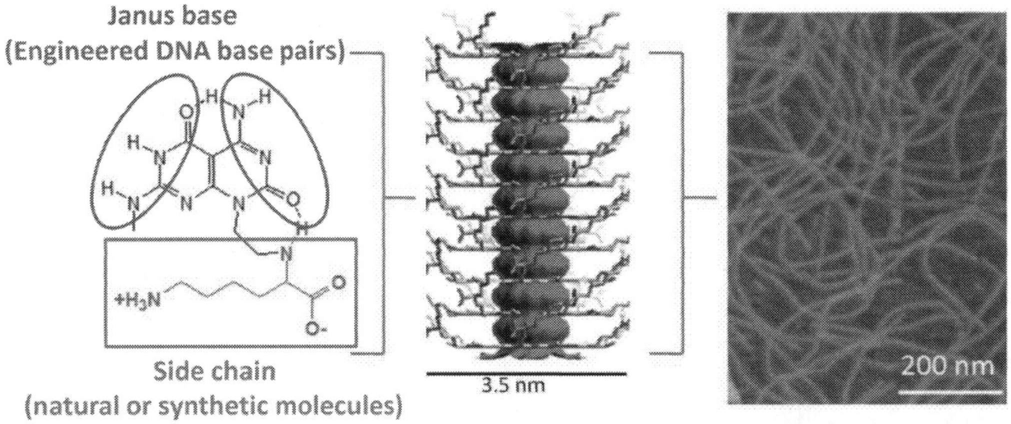

History of Janus Molecules

- Millions of years ago: Janus kinases

They possess two near-identical phosphate-transferring domains. One domain exhibits the kinase activity, while the other negatively regulates the kinase activity of the first.

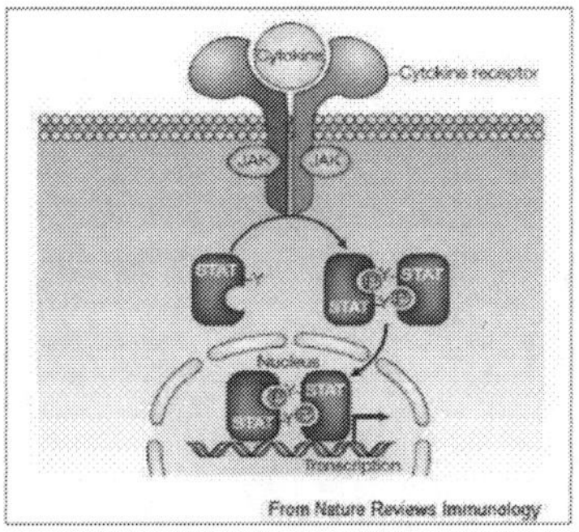

From Nature Reviews Immunology

- In 1980s: Janus-Wedge molecules

For example, Janus molecules to form DNA/RNA triplex (PNA) for sequence-specific gene silence or DNA/RNA-protein interaction.

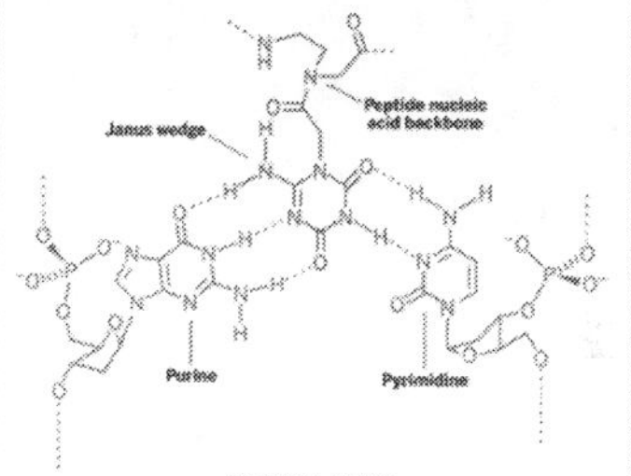

INTERLOPER
Janus wedge residues insert between purine and pyrimidine base pairs in DNA and expropriate their hydrogen bonds.

History of Janus Molecules

- In 1990s: 2D Janus planes

Whitesides 1992: crinkled tape or rosette

Turkenburg 1999: rosette nano-plate

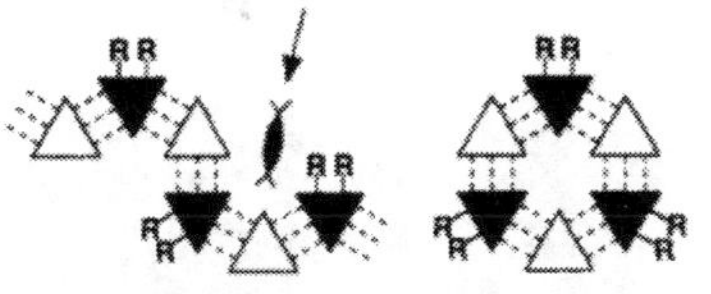
4 crinkled tape **5 rosette**

- After 2000: 3D Janus base nanotubes

Fenniri 2001: GC helical rosette nanotubes

Yu and Chen 2010s: AT and aAT JBNTs

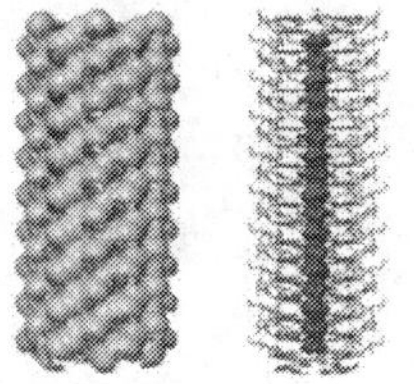

- Today: We have developed "4D" Janus nano-devices for medical applications

Overview of RNAi delivery

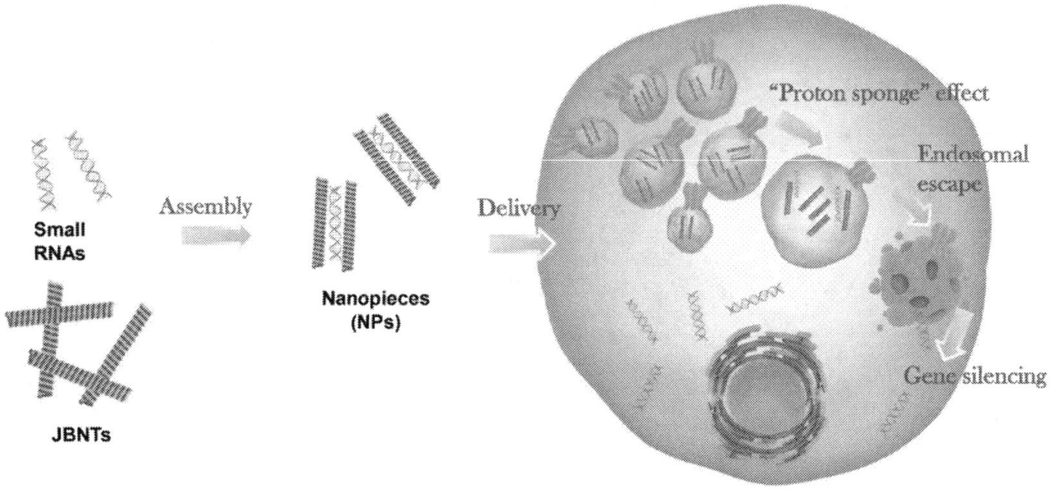

Assembly of NPs by sandwiching RNA among nanotubes via non-covalent interactions

Scale bar: 100nm

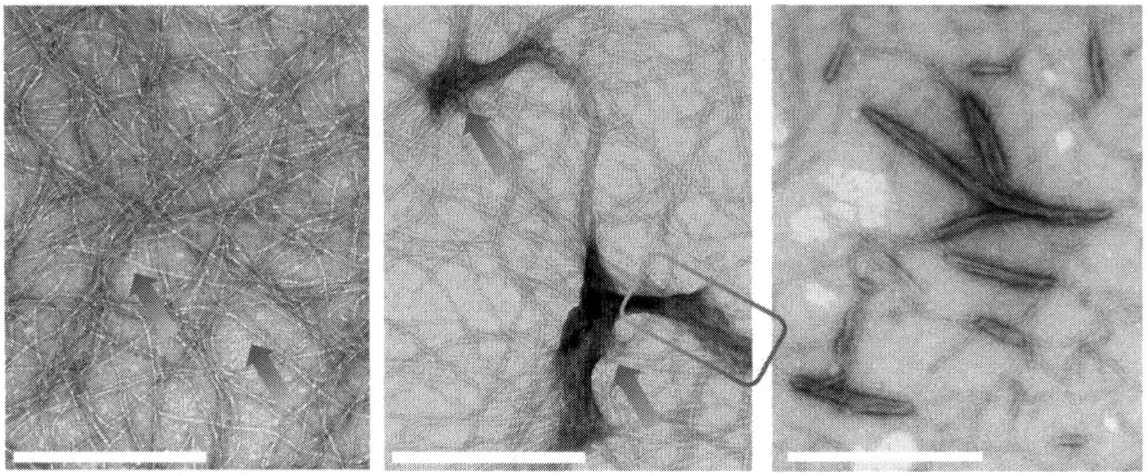

JBNTs	**JBNT/RNA**	**Nanopices (NPs)**
"net-work" morphology	"clots" and "bundles" were formed	after 60s sonication

Assembly of NPs by sandwiching RNA among nanotubes via non-covalent interactions

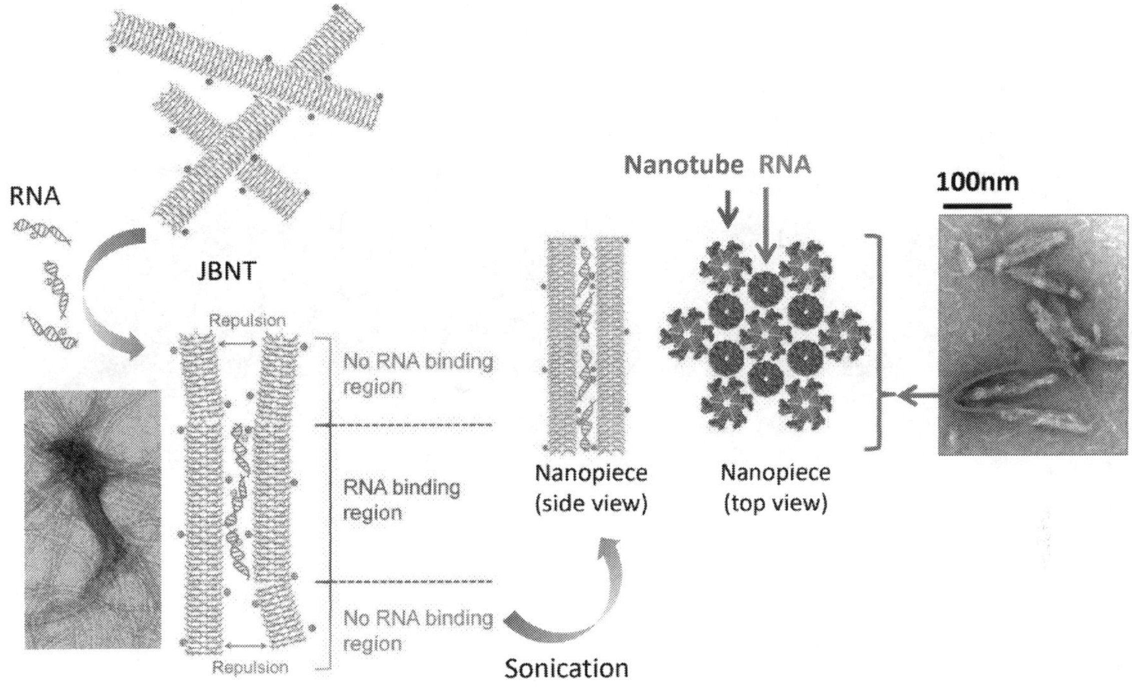

Rapid and efficient siRNA delivery into cells

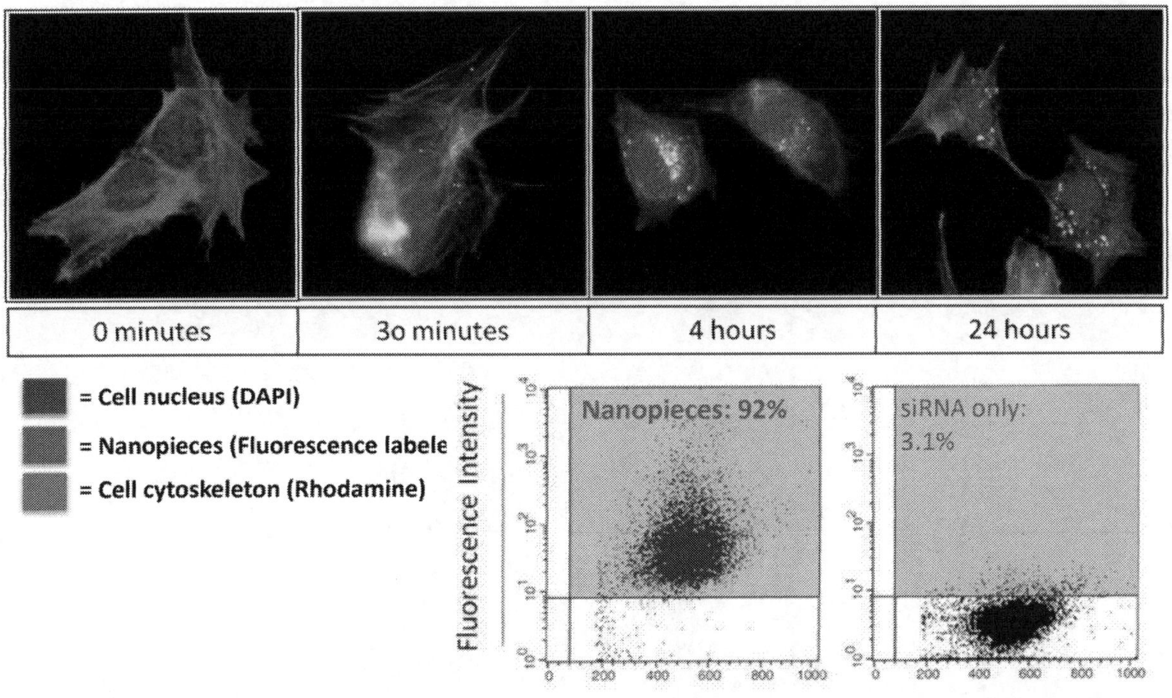

Entering cells via macropinocytosis (same mechanism as liposomes)

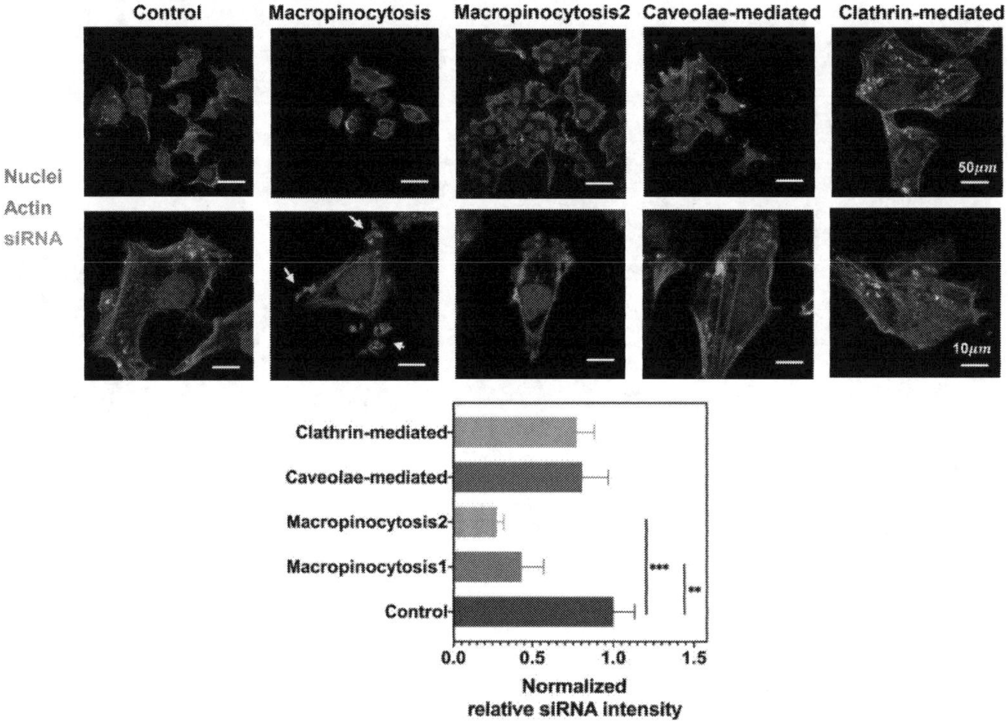

What happens after entering cells?

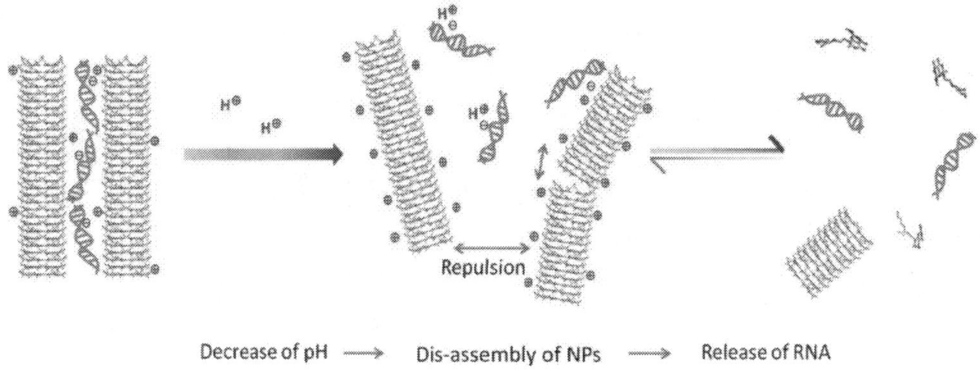

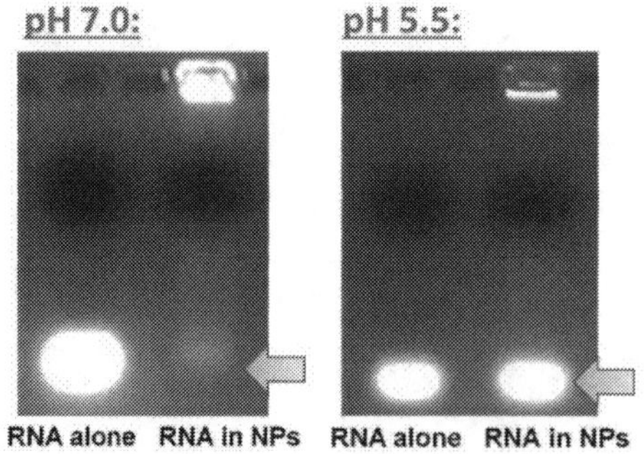

JBNTs for endosomal escape (superior than liposomes)

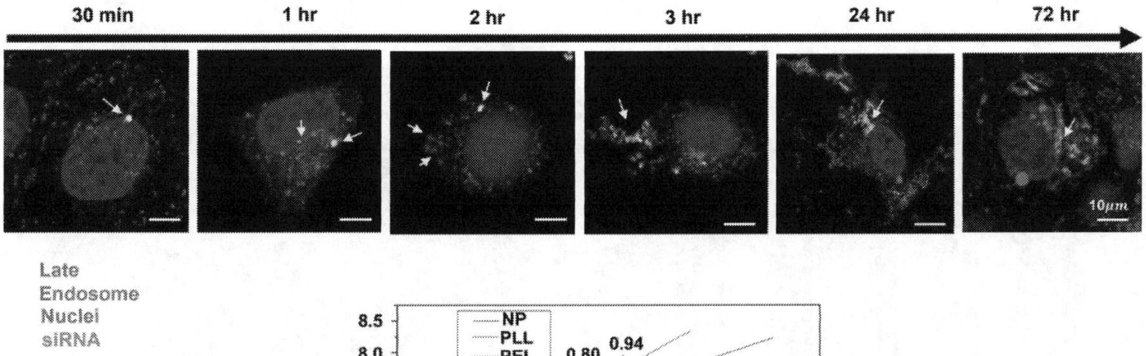

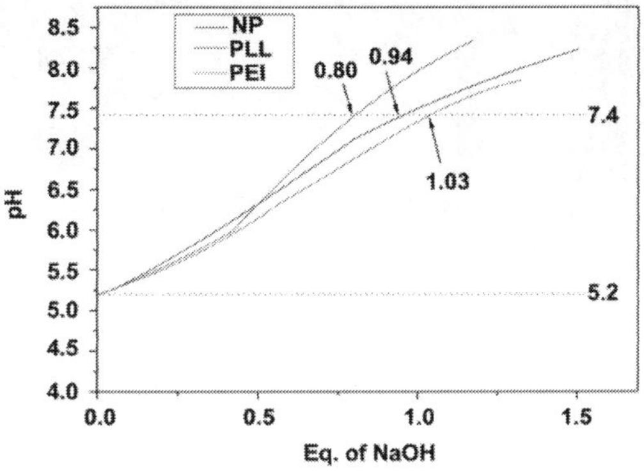

JBNTs for endosomal escape (superior than liposomes)

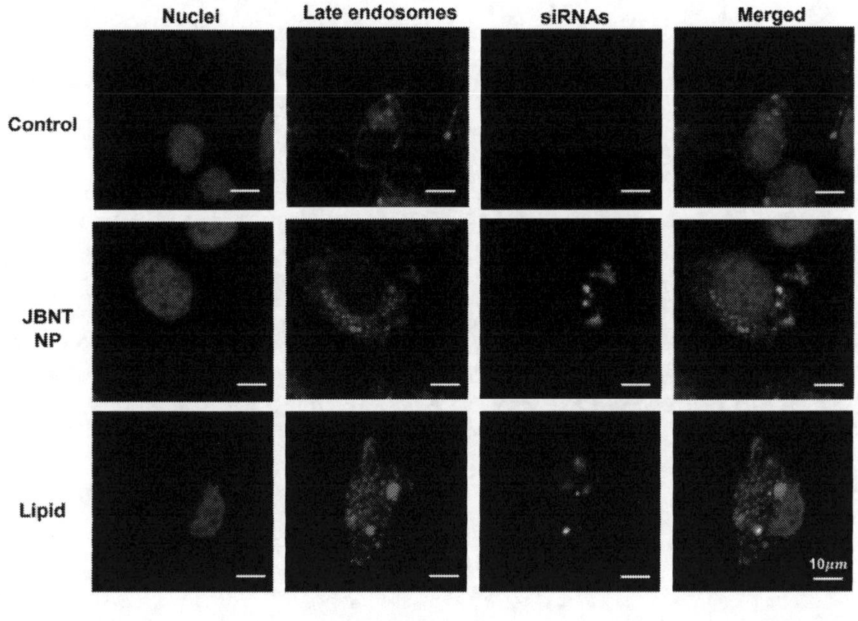

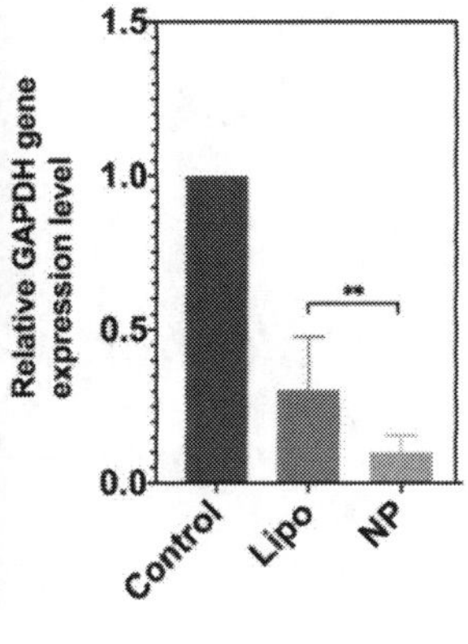

JBNTs present excellent biocompatibility (superior than polymers, lipids and carbon nanotubes)

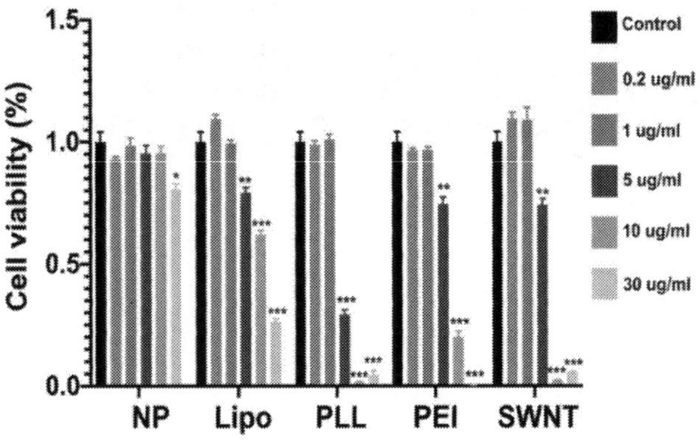

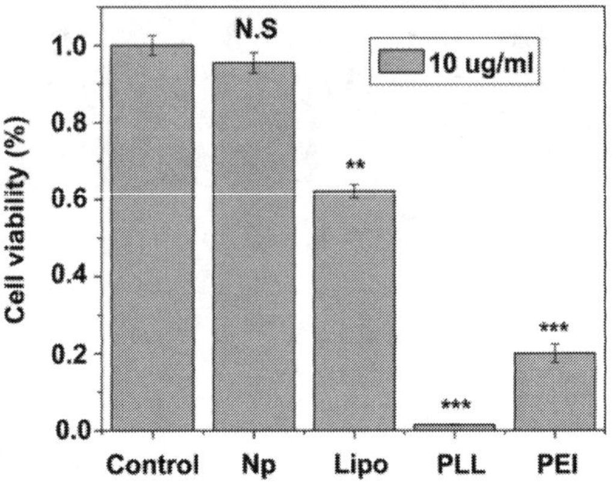

Application:

Delivery into "difficult-to-penetrate" tissues

Example:

Treatment of osteoarthritis (OA)

Motivation: How can we treat OA?

- OA is the leading cause of disability for aged patients. Post-traumatic OA, caused by joint injury, affects 5.6 million Americans.

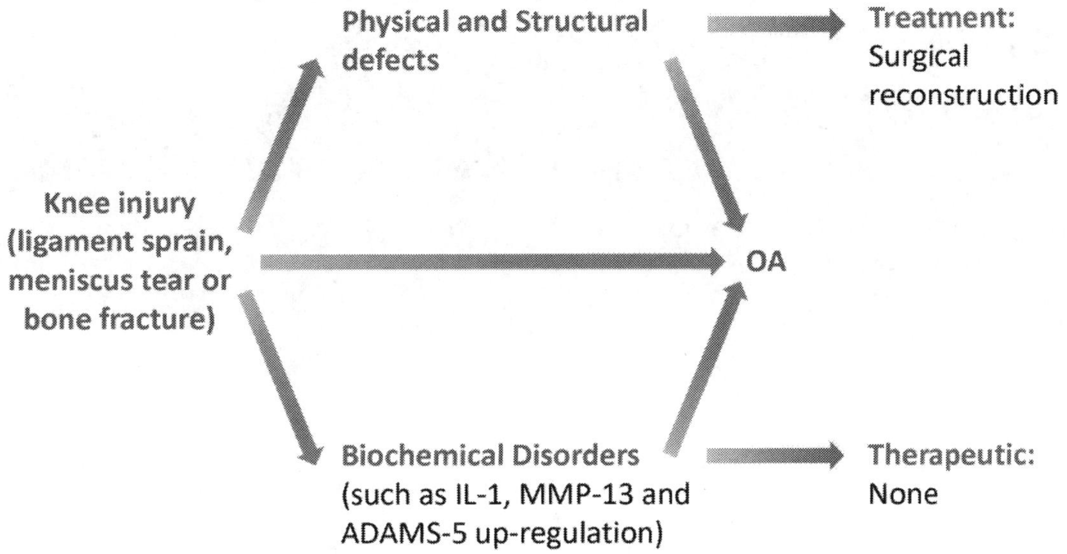

Motivation: How can we treat OA?

- Currently, there is no clinical-approved therapeutic for OA. Corticosteroid and lubrication injections can only temporally relief symptoms (such as pain and movement limitation).
- For example, hyaluronic acid (HA) is a viscous hydrogel. Current commercial HA includes: Euflexxa®, Hyalgan®, Orthovisc®, Supartz®, Synvisc®. However, it does not result in pathological improvement in preventing OA progression, so it was approved by FDA as a medical device but not a drug. Even so, HA has a global market of $5.3 billion in 2012.

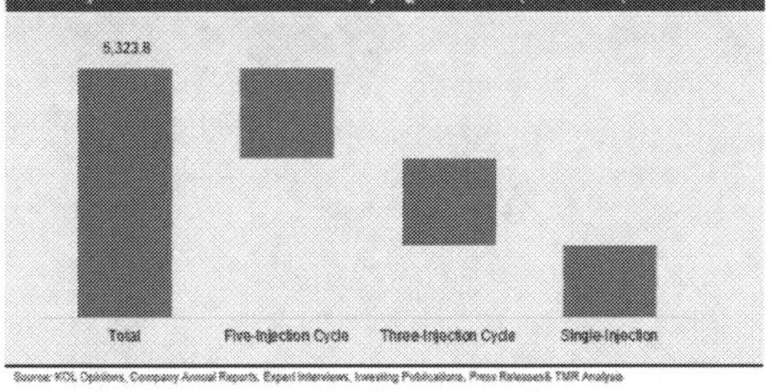

We proposed that the NPs can deliver IL-1R siRNA to penetrate into cartilage and inhibit OA progression.

Intra-cartilage delivery using JBNT NPs

- Nanopiece successful delivered into chondrocytes in mouse cartilage tissue resulting **red fluorescence**.

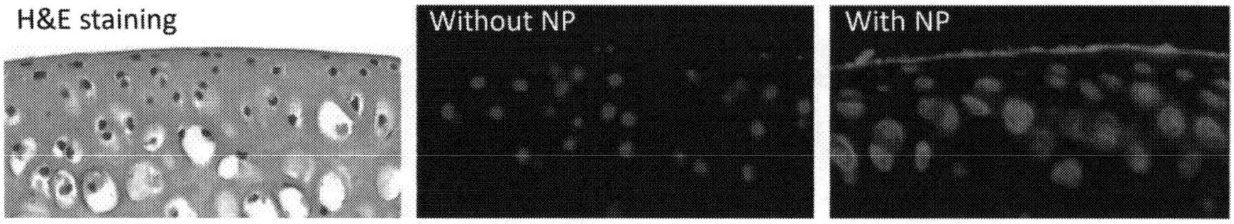

- NPs delivered siRNA into mouse articular cartilage and caused **significant gene knock-down**; while lipofectamine 2000® failed to deliver siRNA into cartilage.

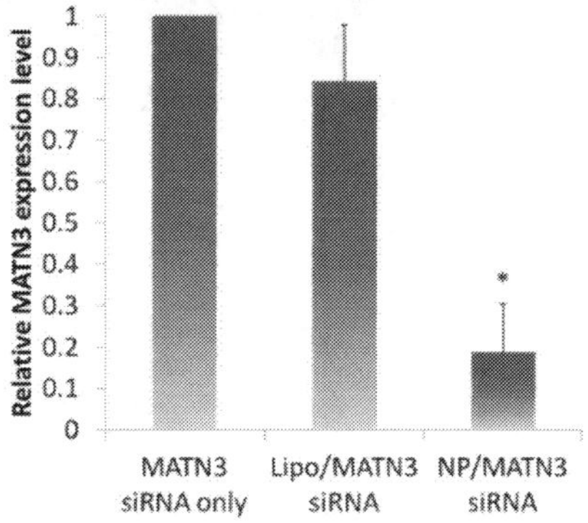

JBNT NPs present a long retention time on cartilage

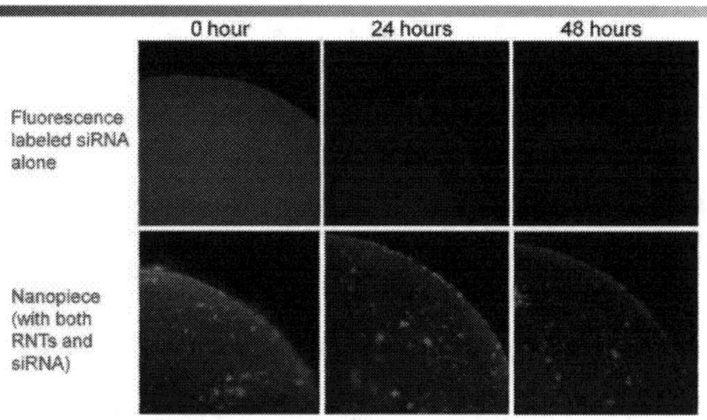

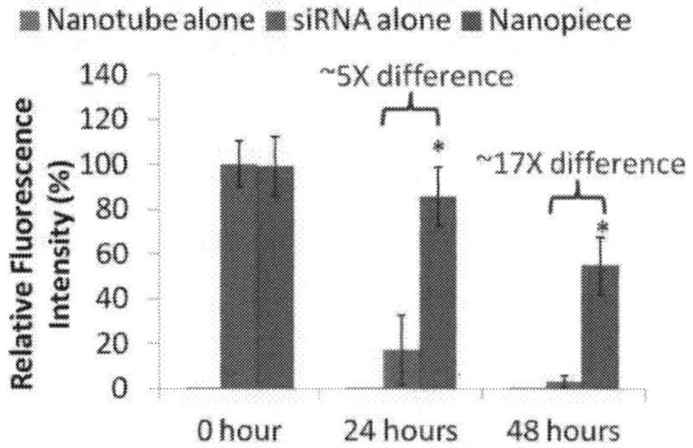

JBNT NPs present a 7 day half-life in vivo (siRNA alone only has a half-life less than 1 day)

NPs:

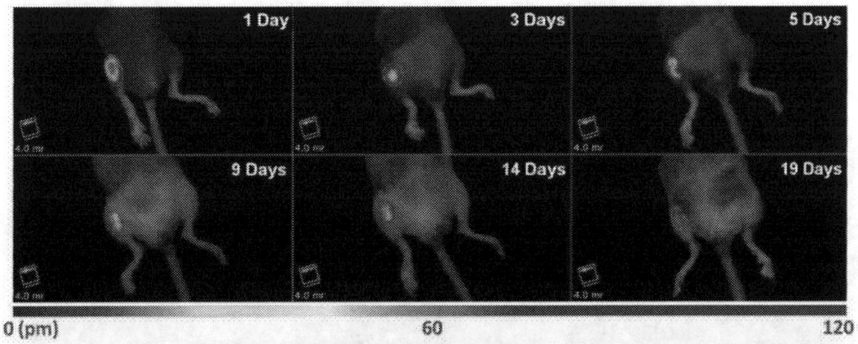

siRNA:

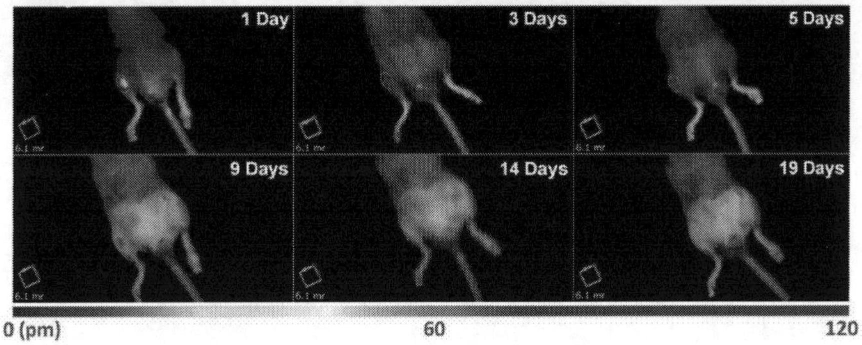

JBNT NPs present a 7 day half-life in vivo (siRNA alone only has a half-life less than 1 day)

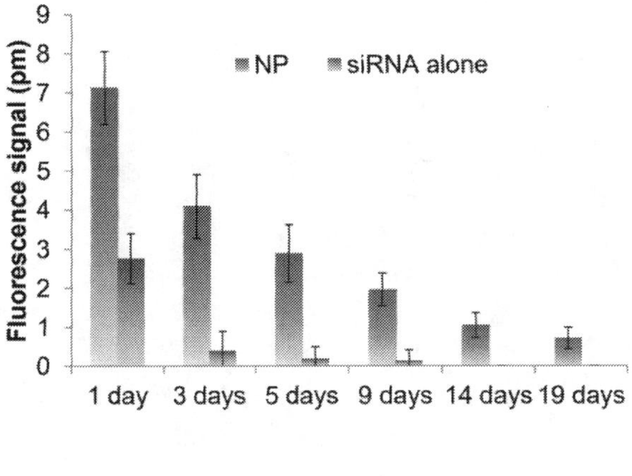

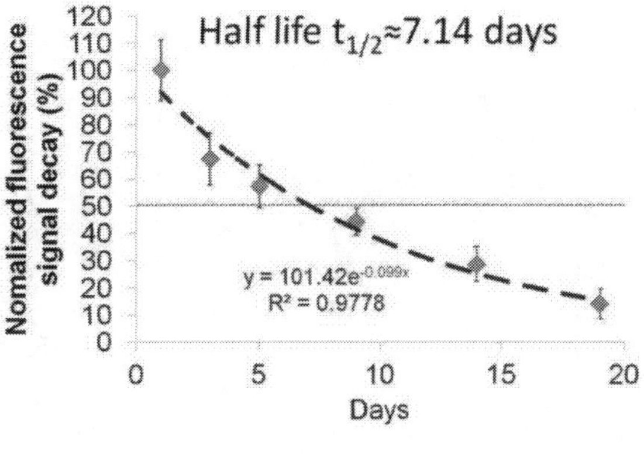

Treatment in an animal model

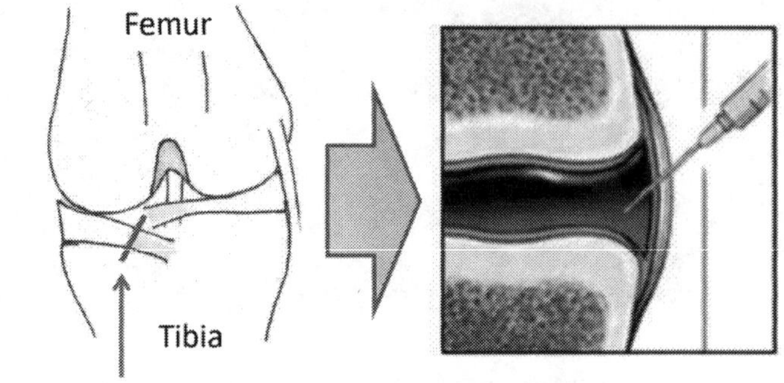

Destabilization of medial meniscus (DMM) to induce OA.

Injection of NPs to inhibit an inflammation gene (IL-1R) as a treatment.

Groups	Negative control	Positive control	Treatment group
Surgery	Sham	DMM	DMM
Reagents	Scrambled siRNA/NP	Scrambled siRNA/NP	IL-1R/NP

Significant IL-1R and its downstream MMP-13 and ADAMTS-5 inhibition with NPs

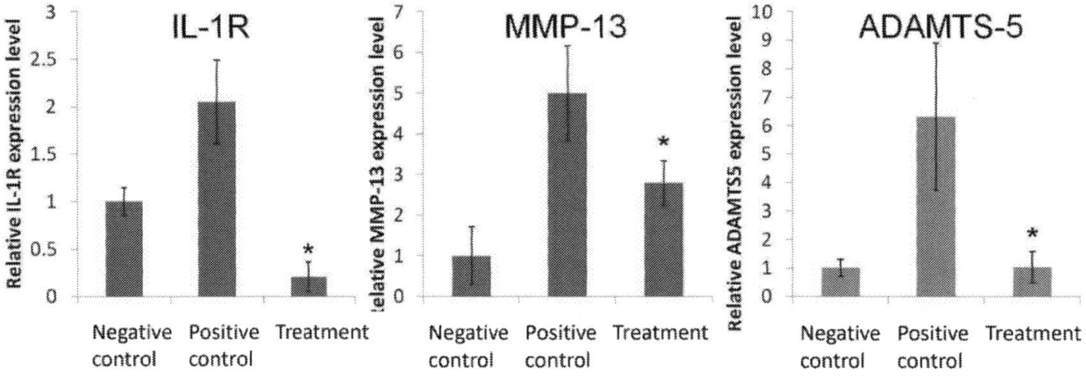

NP treatment dramatically inhibited cartilage degradation in mice

Positive control (without treatment)

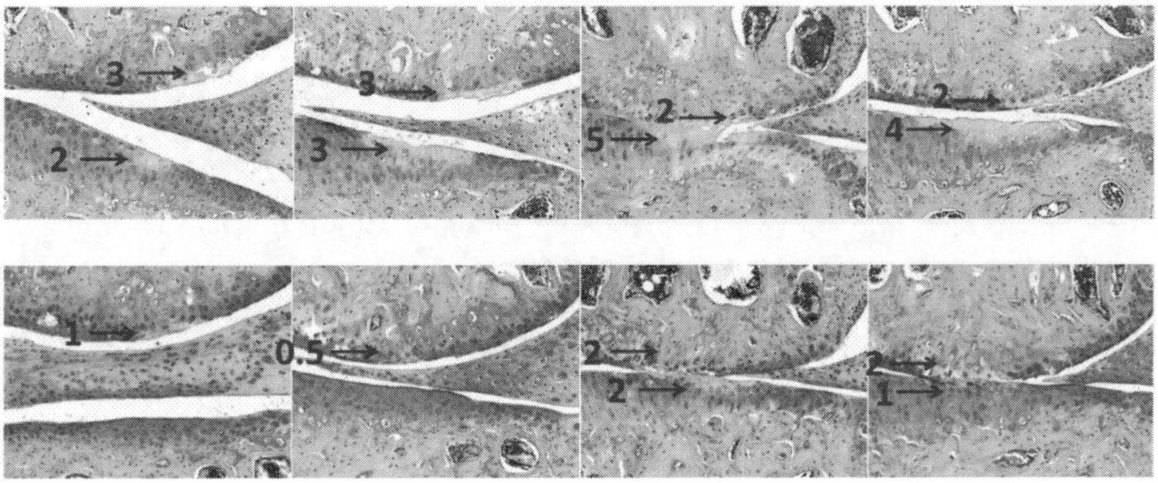

Treatment

IL-1R siRNA/NP greatly inhibited OA progression

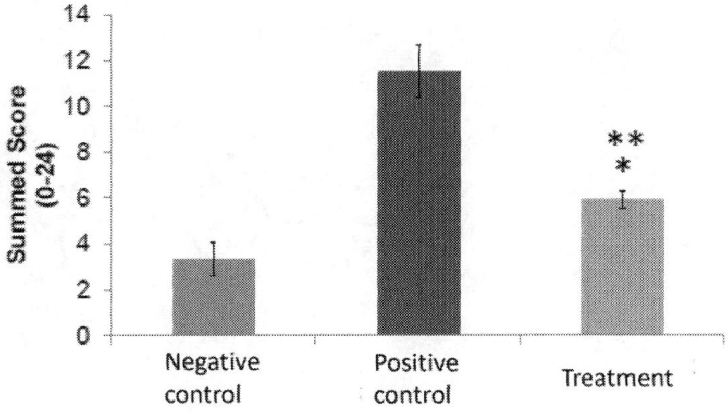

**p<0.01 and *p<0.05 compared with the controls, N=12.

Summary for RNAi Therapy

- The non-covalent delivery vehicles have been constructed based on JBNTs via positive/negative charge interaction and sonication processing.
- JBNT NPs can efficiently enter cells via macropinocytosis (similar to liposomes).
- JBNT NPs can achieve endosomal escape (superior than liposomes).
- JBNT NPs present excellent biocompatibility (superior than cationic polymers, liposomes and carbon nanotubes).
- As an example, NP delivery can deliver into cartilage and significantly inhibit OA progression in an animal model.

Acknowledgements

- Our lab members:

- Funding:

NSF CAREER Award, NIH R01 and NIH R03.

Thank You!

YUPENG CHEN'S
NANOMEDICINE
LABORATORY

611482 Evaluation of Cell Viability and Morphology in Fiber-Hydrogel Composites for Ligament Tissue Engineering

Abstract:

Electrospun fiber scaffolds are of interest for ligament regeneration due to their extracellular matrix-mimicking architecture and high surface area to volume ratio. However, limited cell infiltration into these scaffolds has been attributed to tight fiber packing. Consequently, we are interested in combining electrospun fibers with a space-filling hydrogel phase to decrease fiber packing and improve cell infiltration. To this end, we have fabricated a variety of model composites by combining electrospun polycaprolactone/heparin fibers with a collagen hydrogel. The immediate objective of this study was to determine how construction of the composite affects cell viability and morphology, and the longterm objective is to use these models to study cell migration and tissue formation.

In the first set of studies, we prepared thin layers of electrospun fibers on rigid supports to assess how the placement of cells atop a thin collagen gel coat and beneath 0.5 mm and 1.5 mm of collagen gel affects cell viability and morphology. Results indicated that cells attached to and spread along the supported fibers in all cases, but the thicker (1.5 mm collagen) layer diminished cell metabolic activity relative to the other two configurations. We attribute this to diminished availability of nutrients and oxygen. In the second set of studies, thin layers of electrospun fibers were elevated, and the underside filled with collagen gel (to better mimic the mechanical properties of composites). Cells were then seeded directly atop the fibers or combined with collagen and deposited as a 0.5 mm layer gel layer. For the first case cells were able to attach and spread along fibers, while for the second case some cells were able to find and spread on the fibers, while other cells remained rounded up within the collagen layer. Nevertheless, cells within the entire composite remained viable.

Together, these data show that the thickness of the composite can impact cell metabolic activity while the location of the cells (i.e., on the fibers, in the gel) can influence morphology, both of which we anticipate will affect cell proliferation and ligament tissue formation. As a next step, we are currently studying cell migration in response to the chemoattractant FGF-2. Our next step will be to present ligament-inducing morphogens (e.g., GDF-5) within these fiber-hydrogel composites to evaluate their ability to guide tissue formation.

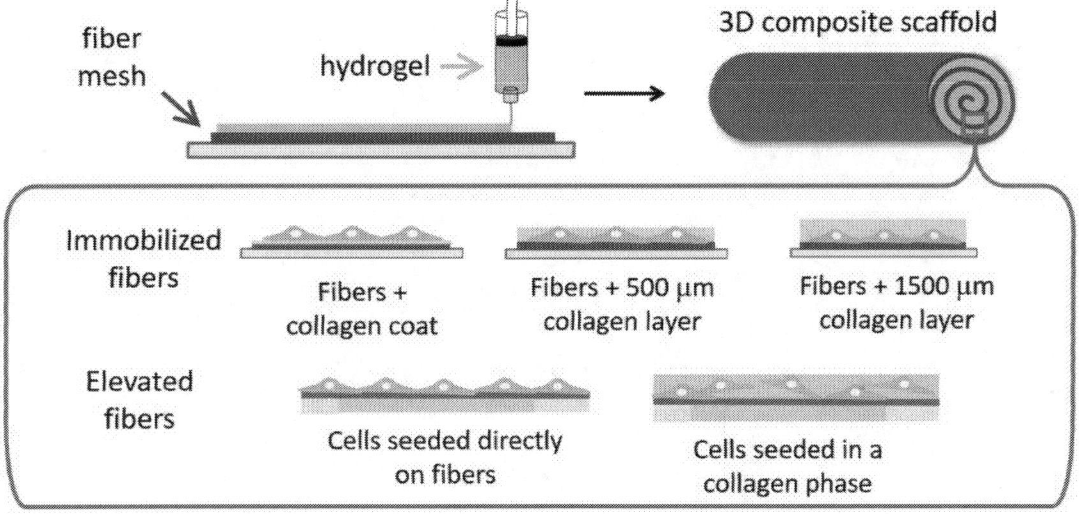

613769 Enhancing the Mechanical and Biological Performance of Bone Scaffolds with Reduced Graphene Oxide

Abstract:

Bone grafting procedures are the second most frequent tissue transplantation, with over two million procedures taking place annually worldwide. There has been a growing demand to develop synthetic bone graft substitutes to overcome the shortage of bone tissue for transplantation. Having expanded into biomedical applications, graphene and its derivatives have attracted significant interest for use in biomaterials due to their unique and remarkable properties. In this work, we developed porous composite scaffolds by incorporating reduced graphene oxide (rGO) into poly(L-lactic-co-glycolic acid) microsphere-based scaffolds, and investigated whether the addition of small amounts of rGO can produce structures that are both mechanically competent and biologically active for bone regenerative engineering. The scaffolds with varying concentrations of rGO were successfully fabricated *via* the oil-in-water emulsion method. The encapsulation and presence of rGO within the polymeric microspheres was confirmed and the physicochemical and microstructural properties of the matrices were evaluated. The addition of rGO, in the moderate range, significantly improved the compressive strength and stiffness of the composite matrices; however, at higher concentrations there was excessive aggregation and re-stacking of the sheets resulting in deterioration of the mechanical properties. All rGO concentrations were well-tolerated by stem cells *in vitro* and none of the scaffolds elicited any cytotoxic effects. Importantly, the moderate ranges of rGO were shown to promote stem cell osteogenic differentiation, corroborating findings that graphene-based materials may exhibit osteoinstructive effects. Furthermore, an *in vivo* study was performed to evaluate the bone regenerative potential and biomechanical competency of these scaffolds in a in a critical-sized load-bearing bone defect model in New Zealand White rabbits.

613765 Recombinant Biomaterials for Treatment of Spinal Cord Injuries

Abstract:

Approximately 15,000 new spinal cord injuries (SCI) occur in the US each year, primarily affecting young adults. Schwann cells are a promising therapy for SCI and are currently being explored in clinical trials; however, significant limitations in cell delivery and long-term survival decrease their therapeutic potential. The low cell retention post-transplantation is partly attributed to (i) mechanical forces during injection that damage the cell membrane and (ii) the lack of a three-dimensional (3D) matrix to support cell viability post-injection. We hypothesized that the development of a shear thinning, injectable hydrogel would improve cell viability, engraftment, and regenerative capacity after transplantation. This hydrogel, produced from a blend of engineered recombinant proteins and peptide-modified synthetic polymer is termed SHIELD: Shear-thinning Hydrogel for Injectable Encapsulation and Long-term Delivery. SHIELD formulations with storage moduli, *G'*, spanning 10-500 Pa all showed excellent viability of encapsulated Schwann cells (>98%) and significant mechanical protection from membrane damage when exposed to syringe needle flow. After 7 days *in vitro*, 3D cultures of within all SHIELD formulations showed positive immunostaining for Schwann cell markers (p75, S100), but cultures in formulations of intermediate stiffness showed higher proliferation rates and decreased caspase activity. In a rat cervical contusion model of SCI, Schwann cells delivered in SHIELD resulted in smaller lesion volumes and improved functional outcomes compared to cells delivered in saline and injury control groups. Since even a mild functional recovery would mean a vast quality-of-life improvement for SCI patients, developing a regenerative therapy for SCI would be extremely significant clinically.

613798 Application of Engineering Tools to Instruct the Regenerative Potential of Cellular Spheroids

Abstract:

Cell therapy provides a versatile and personalized option to address tissue damage due to trauma, inefficient wound repair, degenerative diseases, and cancer. Current cell therapies commonly seek to stimulate tissue regeneration or replace systemic or local deficiencies in cell number or cell function. Cells are generally transplanted as individual, monodispersed cells within a suspension, most commonly by local injection. This strategy has translated to limited success due to rapid cell death upon delivery to the harsh microenvironment, insufficient retention at the site, or even damage due to shear forces associated with injection, shortening their intended effects to a few days. To increase the therapeutic potential of cell-based therapies, we and others have studied the utility of spheroids, which are dense, cellular structures formed into aggregates. However, there remains a tremendous need to develop and apply effective methods for regulating the behavior of cells within spheroids. As synthesis and instruction of spheroids becomes more precise and efficient, their clinical use will emerge as an effective treatment option for regeneration, wound healing, and individualized disease modeling. In this presentation, I will highlight recent and ongoing efforts in our laboratory to utilize engineering tools such as statistical optimization, natural and synthetic biomaterials, and in vitro and in vivo models to regulate and interrogate cell behavior within spheroids.

613723 Engineering Protein-Based Biomaterials for Stem Cell Differentiation

Abstract:

For tissue engineering applications, our lab is developing protein-based biomaterials to promote mesenchymal stem cell differentiation towards particular cell fates (e.g., bone or cartilage). We have genetically engineered modular proteins containing peptides containing growth factors to promote bone differentiation. We recently developed hydrogels containing a blend of collagens I and II, and these hydrogels promote in vitro cartilage differentiation and in vivo cartilage repair. Overall, our work demonstrates that bioactive cues based on peptides or proteins can successfully promote adult stem cell differentiation to desired lineages.

613818 Unconventional Biomaterials for Regenerative Engineering

Abstract:

Introduction: Across various areas of regenerative engineering, there have been limitations with controlling cell adhesion, stem cell differentiation, viability, growth, and blood-material compatibility. Conversion of simple and abundant items to advanced cell culture substrates addresses some of the current challenges in regenerative engineering. The departure from complex fabrication processes to the applications of unusual and ubiquitous materials provides another dimension to the engineering of viable multiscale tissue constructs and regenerative technologies. In addition, the use of sustainable constructs and novel fabrication strategies has the potential to increase access to regenerative engineering technologies on a global scale. In this work, we used unconventional biomaterials such as eggshells and paper for tissue repair.

Materials and Methods: We fabricated eggshell micro/nanoparticle (ESP) reinforced gelatin-based hydrogels to obtain mechanically stable and biologically active three-dimensional (3D) constructs that can differentiate stem cells into osteoblasts. The ESP-reinforced gels were then subcutaneously implanted in a rat model to determine their biocompatibility and degradation behaviors. In addition, these composite scaffolds were used to regenerate critical sized cranial defects in a rat model. We also used mineralized paper scaffolds through an origami-inspired approach to test their osteoinductivity and potential for tissue repair in *in vitro* and *in vivo* studies.

Results and Discussion: The ESP-reinforced scaffolds enabled the differentiation of stem cells without the use of specialized osteogenic growth medium. The ESP-reinforced gels exhibited significant enhancement in mineralization by the cells. Our findings indicated that the ESP composites exhibited superior mechanical properties and showed a favorable *in vivo* response by subcutaneous implantation in a rat model. These scaffolds were highly responsive to cells, and did not elicit inflammatory responses *in vivo*. The implants were easily accepted by the host, allowed for cellular infiltration in 3D, and highly vascularized. Implantation of ESP-reinforced scaffolds into critical sized cranial defects in a rat model resulted in significant bone regeneration in 12 weeks. The resulting bone volume and bone density were as high as the native bone using these composite scaffolds as determined by micro-computed tomography analyses.

We also fabricated origami-inspired paper-based scaffolds for biomineralization. Material properties of the paper-based mineralized scaffolds were determined to be highly tunable. The tensile modulus of the scaffolds increased significantly after the mineralization process. Gene expression results for the osteogenic differentiation markers revealed the osteoinductivity of the mineralized paper scaffolds. Subcutaneous implantation of the samples in rats demonstrated biocompatibility, vascularization, and integration *in vivo*.

Conclusions: Unconventional scaffolds that are readily available and adapted from nature exhibited biomimetic characteristics including porosity, structure, and bioactivity resulting in physiologically relevant constructs. The use of existing naturally derived materials for regenerative engineering provides an inexpensive and sustainable approach that benefits the economy and environment while providing unique solutions to unmet clinical needs. Many of the unconventional biomaterials are overlooked and under-studied for biomedical applications, partially for their simplicity as mundane items.

References: [1]Nguyen, M.A., Camci-Unal, G., Trends in Biotechnology, 38(2): 178-190, 2020. [2]Suvarnapathaki, S., Wu, X., Lantigua, D., Nguyen, M.A., Camci-Unal, G., Nature Asia Materials, 11(1): 1-18, 2019. [3]Wu, X., Stroll, S.I., Lantigua, D., Suvarnapathaki, S., Camci-Unal, G., Biomaterials Science, 7, 2675-2685, 2019.

614320 From Ziplining to Lego Building: Novel Hydrogel Design As Stem Cell Niche to Enhance Tissue Regeneration

Abstract:

Hydrogels are attractive choices of biomaterials for serving as artificial 3D cell niche given their injectability, ease for minimally invasive delivery, as well as tunable chemical and physical properties. In this talk, I will share two examples of our recent work on developing novel hydrogels with unique properties to enhance stem cell differentiation and musculoskeletal tissue regeneration. Using physically-crosslinked alginate hydrogels, it has been reported that molecular mobility combined with viscoelasticity modulates stem cell differentiation. However, one key limitation of the existing hydrogel tools is the inability to decouple changes in viscoelasticity from molecular mobility. As such, how tuning molecular mobility alone modulate stem cell fates in 3D remains unknown. In the first example, I will share our work on developing novel sliding hydrogels that allows crosslinks and biochemical ligands to slide along the hydrogel backbone, enabling encapsulating cells to "zipline" crosslinks and ligands in 3D with independently tunable molecular mobility. Our results validate molecular mobility as a novel parameter in biomaterials design to accelerate stem cell differentiation towards multiple lineages, with improved tissue structures and functions. In the second example, I will share a "lego-like", microribbon (μRB)-based hydrogels platform recently invented by our group. Conventional hydrogels are often mechanically weak, and lack macroporosity desirable for cell and tissue growth. To overcome these limitations, we developed a novel two-step fabrication process that involves first wet-spinning hydrogels into μRB-like structures, which can further photocrosslink into a 3D macroporous scaffolds. The combination of injectability and macroporosity is especially useful for repairing large tissue defects with clinically relevant dimensions in a minimally invasive manner. The biochemical and physical properties of μRBs can be independently tuned to promote desirable cell phenotype and tissue responses. We further developed in situ drug delivery components that support tunable drug delivery from these μRB scaffolds to promote tissue regeneration in vivo. Compared to conventional hydrogels, μRB-based hydrogels substantially accelerate MSC differentiation and new tissue deposition in vitro and in vivo for repairing various tissues including cartilage and bone with much faster restoration of tissue load-bearing functions. In summary, the sliding hydrogels and μRB-based hydrogels offer novel injectable biomaterials for cell delivery, and are attractive scaffold choices to enhance regeneration of various tissue types.

Acknowledgements: The authors would like to thank NIH R01DE024772, NIH 1R01AR074502, Stanford coulter translational grant, Stanford bio-X IIP seed grant, and Stanford SPARK program for funding.

AUTHOR INDEX

AIChE
120 Wall Street, FL 23
New York, NY 10005-4020

ISBN 978-1-7138-2286-8
9 781713 822868

CURRAN ASSOCIATES INC.
proceedings.com